土星

群星环绕的奇迹星球

SATURN

The Ringed Wonder

（英国）埃伦·劳伦斯／著　　张　骁／译

江苏凤凰美术出版社

著作权合同登记图字：10-2022-144

图书在版编目（CIP）数据

土星：群星环绕的奇迹星球 / （英）埃伦·劳伦斯

著；张骁译 . -- 南京：江苏凤凰美术出版社，2025.

1. -- （环游太空）. -- ISBN 978-7-5741-2027-3

Ⅰ . P185.5-49

中国国家版本馆 CIP 数据核字第 2024RC8866 号

策　　　　划　朱婧
责　任　编　辑　高　静　奚　鑫
责　任　校　对　王　璇
责任设计编辑　樊旭颖
责　任　监　印　生　嫄
英　文　朗　读　C.A.Scully
项　目　协　助　邵楚楚　乔一文雯

丛　书　名　　环游太空
书　　　名　　土星：群星环绕的奇迹星球
著　　　者　　（英国）埃伦·劳伦斯
译　　　者　　张　骁
出　版　发　行　江苏凤凰美术出版社（南京市湖南路 1 号　邮编：210009）
印　　　刷　　南京新世纪联盟印务有限公司
开　　　本　　710 mm×1000 mm　1/16
总　印　张　　18
版　　　次　　2025 年 1 月第 1 版
印　　　次　　2025 年 1 月第 1 次印刷
标　准　书　号　ISBN 978-7-5741-2027-3
总　定　价　　198.00 元（全 12 册）

目录 Contents

书中加粗的词语见词汇表解释。
Words shown in **bold** in the text are explained in the glossary.

欢迎来到土星
Welcome to Saturn

想象一下，你正在飞往一个离地球超过10亿千米的星球。

Imagine flying to a world that is more than a billion kilometers from Earth.

随着飞船越来越近，你看到了这个星球被巨大圆环围绕着。

As your spacecraft gets close, you see this world is surrounded by huge rings.

这些圆环是由尘埃和冰块构成的。

The rings are made of dust and chunks of ice.

你成功地飞过了这些冰冻的圆环，但是却根本没有地方着陆。

You manage to fly through the icy rings, but there is nowhere to land.

因为这个遥远星球是一个主要由气体和液体组成的巨大球体。

That's because this distant world is a gigantic ball of **gases** and liquids.

欢迎来到行星土星！

Welcome to the **planet** Saturn!

人类从未到达过土星，但是飞行器做到了。飞行器将关于这颗星球的信息传回给地球上的科学家们。

No humans have ever visited Saturn, but spacecraft have. The spacecraft sent information about the planet back to scientists on Earth.

这张图片是由电脑制成的，它展现了飞越土星环时你可能看到的景象。

This picture was created on a computer. It shows what it might look like to fly into the rings that circle around Saturn.

太阳系 The Solar System

土星以将近35 000千米每小时的速度在太空移动。

Saturn is moving through space at about 35,000 kilometers per hour.

它围绕着太阳做一个巨大的圆周运动。

It is moving around the Sun in a huge circle.

土星是围绕太阳公转的八大行星之一。

Saturn is one of eight planets circling the Sun.

八大行星分别是水星、金星、我们的母星地球、火星、木星、土星、天王星和海王星。

The planets are called Mercury, Venus, our home planet Earth, Mars, Jupiter, Saturn, Uranus, and Neptune.

冰冻的彗星和被称为"小行星"的大型岩石也围绕着太阳公转。

Icy **comets** and large rocks, called **asteroids**, are also moving around the Sun.

太阳、行星和其他天体共同组成了"太阳系"。

Together, the Sun, the planets, and other space objects are called the **solar system**.

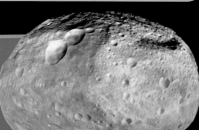

太阳系中的大多数小行星都集中在被称为"小行星带"的环状带中。

Most of the asteroids in the solar system are in a ring called the asteroid belt.

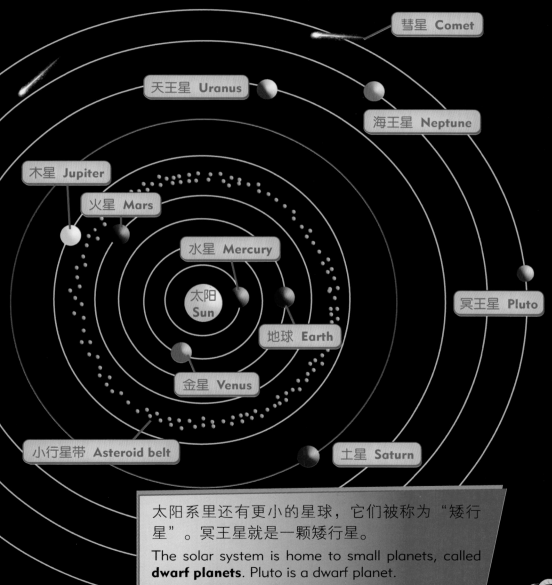

太阳系 **The Solar System**
土星是离太阳第六近的行星。
Saturn is the sixth planet from the Sun.

彗星 **Comet**

天王星 **Uranus**

海王星 **Neptune**

木星 **Jupiter**

火星 **Mars**

水星 **Mercury**

太阳
Sun

冥王星 **Pluto**

地球 **Earth**

金星 **Venus**

小行星带 **Asteroid belt**

土星 **Saturn**

太阳系里还有更小的星球，它们被称为"矮行星"。冥王星就是一颗矮行星。

The solar system is home to small planets, called **dwarf planets**. Pluto is a dwarf planet.

土星的奇幻之旅
Saturn's Amazing Journey

行星围绕太阳公转一圈所需的时间被称为"一年"。

The time it takes a planet to **orbit**, or circle, the Sun once is called its year.

地球绕太阳公转一圈需要略多于365天，所以地球上的一年有365天。

Earth takes just over 365 days to orbit the Sun, so a year on Earth lasts 365 days.

土星比地球离太阳更远，所以它绕太阳公转的路程更长。

Saturn is farther from the Sun than Earth, so it must make a much longer journey.

土星绕太阳公转一圈大约需要30个地球年。

It takes Saturn nearly 30 Earth years to orbit the Sun.

这意味着，地球上的一个30岁的成年人在土星上才刚满1岁！

This means that a 30-year-old adult on Earth would just be turning 1 in Saturn years!

当行星围绕太阳公转时，它也像陀螺一样自转着。

As a planet orbits the Sun, it also spins, or **rotates**, like a top.

土星 Saturn

土星环 Ri

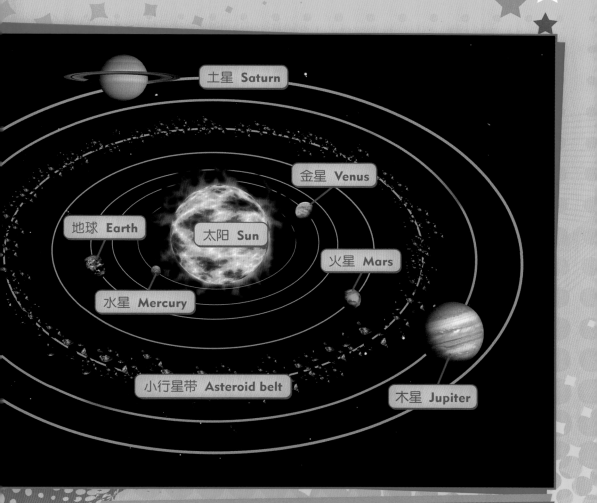

土星 Saturn

金星 Venus

地球 Earth

太阳 Sun

火星 Mars

水星 Mercury

小行星带 Asteroid belt

木星 Jupiter

地球绕太阳公转一圈的路程约为9.4亿千米，而土星绕太阳公转一圈的距离要长得多，约为90亿千米！

To orbit the Sun once, Earth makes a journey of nearly 940 million kilometers. Saturn makes a much longer journey of about 9 billion kilometers!

近距离观察土星
A Closer Look at Saturn

土星是太阳系中第二大行星。

Saturn is the second-largest planet in the solar system.

不同于岩质行星地球，土星没有固体表面。

Unlike Earth, which is a rocky planet, Saturn doesn't have a solid surface.

这颗行星被一层厚厚的气体包围，称为"大气层"。

The planet is surrounded by a thick layer of gases called an **atmosphere**.

在大气层之下，这颗行星其实是一颗由液体组成的巨大球体。

Beneath its atmosphere, the planet is a gigantic ball of liquids.

在土星的大气层内部有着超快的气流，最高能达到1600千米每小时。

Inside Saturn's atmosphere, super-fast winds blow at up to 1,600 kilometers per hour.

土星 Saturn

地球 Earth

在这张图片中，我们把地球和土星放在了一起。现在可以看出，和地球相比，土星究竟有多大了吧。

In this picture, Earth and Saturn have been placed next to each other. The picture shows how huge Saturn is compared to Earth.

这张照片显示了土星北极的一个超大飓风，是由飞行器"卡西尼号"于2013年4月拍摄的。这个红色部分被称为"飓风眼"，也就是飓风的中心。

This photo shows a giant **hurricane** at Saturn's north pole. It was taken by a spacecraft named *Cassini* (kuh-SEE-nee) in April 2013. The red part is the eye, or center, of the hurricane.

飓风眼 **Eye of hurricane**

飓风眼的特写
Close-up of eye of hurricane

这张图片是土星上飓风眼的特写。这个飓风眼的直径达2 000千米！

This is a close-up of the eye of the hurricane on Saturn. The eye is 2,000 km across!

土星的奇妙圆环
Saturn's Fantastic Rings

如果你透过望远镜看土星，会觉得它的环看上去像是一个固态的、色彩斑斓的圆圈。

When you look at Saturn through a telescope, its rings look like solid, colorful circles.

实际上土星环是由无数片冰块和尘埃构成的。

They are actually made of billions of pieces of ice and dust.

这些圆环可能是由靠近土星的卫星和彗星的碎片形成的。

The rings may have formed from pieces of **moons** and comets that came close to Saturn.

这些碎片一边绕着土星运动，一边发生剧烈的互相撞击。

As these objects orbited the planet, they smashed into one another.

这些撞击形成了大量的冰块和尘埃。

This created lots of chunks of ice and dust.

亿万年后，这些冰块碎片就聚合起来，共同形成了土星环。

Over millions and billions of years, this icy rubble gathered together to form the rings.

在土星环中，有些碎冰和尘埃就像沙粒一样小，有些像房子一样大，还有少数的一些能像一座山一样大！

Some of the pieces of ice and dust in Saturn's rings are as small as grains of sand. Others are as big as houses, while a few are the size of mountains!

土星环 **Rings**

土星 **Saturn**

土星的卫星家族
Saturn's Family of Moons

有许多体积较小的星球绕着土星旋转，它们组成了一个规模庞大的家族。

Saturn has a big family of small worlds circling around it.

这些覆满冰的星体都是这颗庞大星球的卫星。

These icy space objects are the huge planet's moons.

我们的家园——地球，只有一颗卫星。

Earth, our home planet, has just one moon.

已发现土星至少有146颗卫星，而且科学家们认为也许还有更多！

Saturn has at least 146 moons, and scientists think there may be more!

土星最大的卫星——泰坦（土卫六）是太阳系中第二大卫星。

Saturn's biggest moon, Titan, is the second-largest moon in the solar system.

它也是太阳系中唯一一颗有大气层的卫星。

It is also the only moon in the solar system with an atmosphere.

泰坦比地球的卫星要大得多。
Titan is much bigger than Earth's moon.

地球的卫星 Earth's moon

泰坦 Titan

地球 Earth

斯普莱特
The Splat

利亚（土卫五）　**Rhea (REE-uh)**

土星许多卫星的表面都有一些坑坑洼洼的洞，被叫作陨石坑。这些陨石坑是太空中的物体与卫星碰撞留下的痕迹。土星的卫星利亚（土卫五）上有一个陨石坑，名为斯普莱特。

Many of Saturn's moons have holes called craters on their surfaces. The craters were made by space objects hitting the moons. Saturn's moon Rhea has a crater known as "The Splat".

埃庇米修斯
（土卫十一）

**Epimetheus
(eh-pih-MEE-thee-us)**

米玛斯
（土卫一）

**Mimas
(MI-muhs)**

海伯利安
（土卫七）

**Hyperion
(hi-PEER-ee-uhn)**

恩克拉多斯
（土卫二）

**Enceladus
(en-SEL-eh-duhs)**

菲比
（土卫九）

**Phoebe
(FEE-bee)**

伊阿帕托斯
（土卫八）

**Iapetus
(i-AP-eh-tuhs)**

15

探测土星与泰坦的任务
A Mission to Saturn and Titan

1997年10月，一个名为"卡西尼号"的飞行器从地球发射升空，它的任务是去探测土星。

In October 1997, a spacecraft named *Cassini* blasted off from Earth on a mission to study Saturn.

这个飞行器于2004年7月到达土星。

It reached Saturn in July 2004.

为了研究土星的卫星泰坦，"卡西尼号"还携带了一个名为"惠更斯号"的小型空间探测器。

Cassini sent a small space **probe**, called *Huygens* (HOY-guhnz), to study Saturn's moon Titan.

"惠更斯号"在穿越泰坦的大气层时，拍摄了不少照片，收集到了十分珍贵的数据。

As *Huygens* traveled through Titan's atmosphere, it took pictures and collected **data**.

当它在卫星上着陆的时候，它也收集了大量信息。

It also collected information when it landed on the moon.

"惠更斯号"将它收集到的数据传回"卡西尼号"用了大约90分钟。

Huygens beamed the data it gathered back to *Cassini* for about 90 minutes.

这张照片显示了"卡西尼号"环绕土星运动的样子。

This picture shows how *Cassini* looked as it orbited Saturn.

"卡西尼号" *Cassini*

土星 Saturn

泰坦上的山脉
A mountain on Titan

"惠更斯号"让科学家们能够看到泰坦大气层下的模样。它拍摄到了山脉、湖泊和岛屿的图像。这张照片是"惠更斯号"在距泰坦地表8 000米的地方拍摄的。

Huygens allowed scientists to see what was under Titan's atmosphere. It captured images of mountains, lakes, and islands. This photo was taken when *Huygens* was about 8 km above the surface.

"惠更斯号"探测器
Huygens probe

这张图片展示了"惠更斯号"循序渐进地降落在泰坦上的样子。这架探测器大概跟一辆小轿车一样大。

This step-by-step picture shows how *Huygens* might have looked as it landed on Titan. The probe was about the size of a car.

奇妙的探测任务！
An Amazing Mission!

到2017年9月为止，"卡西尼号"一直向地球传输着土星及其卫星的数据。

Cassini sent data about Saturn and its moons back to Earth until September 2017.

科学家们将会花很多年来研究这些信息和图像。

It will take scientists many years to study all the information and images.

"卡西尼"号发现了一个环绕土星的新土星环，还有8颗新的卫星。

Cassini discovered a new ring circling Saturn and eight new moons.

它在恩克拉多斯（土卫二）上还发现了水、气体和一些有趣的化学反应。

It also discovered water, gases, and some interesting chemical activity on Saturn's moon, Enceladus.

科学家认为，这些发现意味着在恩克拉多斯上可能生活着一些渺小的微生物。

Scientists think these discoveries mean that tiny living things called **microbes** could be alive on Enceladus.

甚至在土星的其他卫星上也可能有这样的微生物！

There might even be microbes on some of Saturn's other moons, too!

当任务结束的时候，"卡西尼号"飞向了土星环，并且在行星的大气层中燃烧殆尽。为什么科学家要焚毁这个飞行器呢？是为了保证它不会撞到可能有生物存在的卫星。

When its mission ended, *Cassini* flew through Saturn's rings and burned up in the planet's atmosphere. Why did scientists destroy the probe? To be sure that *Cassini* could never hit one of the moons where there might be living things.

这张照片是由"卡西尼号"拍摄的，它显示了土星面对太阳时的状态。在土星的背面，我们可以看见这颗行星的巨大阴影。

This photo was taken by *Cassini*. It shows Saturn facing the Sun. The planet's huge shadow can be seen behind it.

土星的暗面 **Saturn's shadow**

土星 **Saturn**

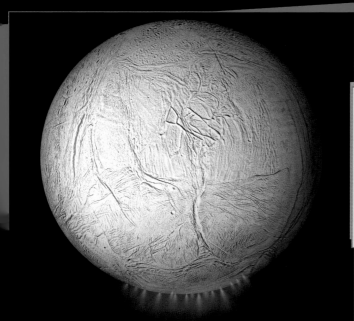

这幅图片显示了冰和气体是如何从恩克拉多斯地表的裂缝中喷涌而出的。

This illustration shows how ice and gases spray from cracks in the surface of Enceladus.

有趣的土星知识
Saturn Fact File

以下是一些有趣的土星知识：土星是距离太阳第六近的行星。

Here are some key facts about Saturn, the sixth planet from the Sun.

土星的发现
Discovery of Saturn

在地球上即便不使用望远镜，土星也可以被观测到。早在古代，人们就已经知道了土星的存在。

Saturn can be seen in the sky without a telescope. People have known it was there since ancient times.

土星是如何得名的
How Saturn got its name

这颗行星是以古罗马农神的名字命名的。

The planet is named after the Roman god of farming.

行星的大小
Planet sizes

这张图显示了太阳系八大行星的大小对比。

This picture shows the sizes of the solar system's planets compared to each other.

水星 Mercury
地球 Earth
太阳 Sun
木星 Jupiter
金星 Venus
火星 Mars
天王星 Uranus
土星 Saturn
海王星 Neptune

土星的大小
Saturn's size

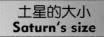

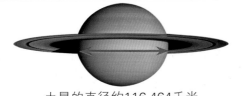

土星的直径约116 464千米

About 116,464 km across

土星自转一圈需要多长时间
How long it takes for Saturn to rotate once

大约10.5个地球时

About 10.5 Earth hours

土星与太阳的距离
Saturn's distance from the Sun

土星与太阳的最短距离是1 349 823 615千米。
土星与太阳的最远距离是1 503 509 229千米。

The closest Saturn gets to the Sun is 1,349,823,615 km.
The farthest Saturn gets from the Sun is 1,503,509,229 km.

土星围绕太阳公转的平均速度
Average speed at which Saturn orbits the Sun

每小时34 701千米
34,701 km/h

土星绕太阳的轨道长度
Length of Saturn's orbit around the Sun

8 957 504 604千米
8,957,504,604 km

太阳
Sun

土星轨道 Saturn's orbit

土星 Saturn

土星上的一年
Length of a year on Saturn

大约11 000个地球天（大约30个地球年）
Nearly 11,000 Earth days (nearly 30 Earth years)

土星的卫星
Saturn's moons

土星至少有146颗卫星。可能还有更多的有待发现。

Saturn has at least 146 moons. There are possibly more to be discovered.

 ## 土星上的温度
Temperature on Saturn

零下178摄氏度

-178°C

动动手吧：闪闪发亮的土星
Get Crafty : Sparkly Saturn

当太阳光照射到土星环上的时候，环上的冰块就会闪闪发亮。用水彩和亮片制作属于你自己的闪闪发亮的土星吧！

你需要：

- 彩色美术纸
- 一个圆口玻璃瓶
- 一支铅笔
- 一把剪刀
- 水彩和水彩刷
- 胶水
- 亮片
- 一把直尺

1. 画出土星：把圆口玻璃瓶倒扣在纸上，沿着瓶口边缘画一个圆形，把这个圆形剪下来。

2. 制作土星环：画一个比圆形宽两倍的椭圆形，然后把椭圆形剪下来。

椭圆形　圆形

3. 给圆形涂色。你会用什么颜色来表现土星的云层与大气层呢？

4. 在椭圆形的一面涂上胶水，把亮片撒在胶水上，直到整个表面都撒满。当胶水干了以后，小心地把没有粘上的亮片抖下，回收。

5. 测量土星的宽度，然后请成年人帮忙，在椭圆形的中间划出一个长度为土星直径的划口。

划口

6. 最后，把土星插入划口中，使它周围环绕着闪闪发光的亮片圆环。

词汇表 Glossary

小行星 | asteroid

围绕太阳公转的大块岩石，有些小得像辆汽车，有些大得像座山。

大气层 | atmosphere

行星、卫星或恒星周围的一层气体。

彗星 | comet

由冰、岩石和尘埃组成的天体，围绕太阳公转。

数据 | data

指各种形式的信息和事实，被收集后可供研究。

矮行星 | dwarf planet

围绕太阳运行的圆形或近圆形天体，比八大行星小得多。

气体 | gas

无固定形状或大小的物质，如氧气或氦气。

飓风 | hurricane

巨大的风暴，环绕风暴中心（风眼）旋转。直径可达数百千米，风速高达每小时320千米。

微生物 | microbe

极小的生物，无法用裸眼看见。能够使人生病的细菌就是一种微生物。

卫星 | moon

围绕行星运行的天体。通常由岩石或岩石和冰构成。直径从几千米到几百千米不等。地球有一个卫星，名为"月球"。

公转 | orbit

围绕另一个天体运行。

行星 | planet

围绕太阳公转的大型天体：一些行星，如地球，主要是由岩石组成的；其他的行星，如木星，主要是由气体和液体组成的。

探测器 | probe

不载人太空飞船。通常被送往行星或其他天体，用于拍摄照片并收集信息，由地球上的科学家操作控制。

自转 | rotate

物体自行旋转的运动。

太阳系 | solar system

太阳和围绕太阳公转的所有天体，如行星及其卫星、小行星和彗星。